Biology's Big Bang: The Cambrian Explosion

Paul K. Chien

SEATTLE   DISCOVERY INSTITUTE PRESS   2024

Description

Darwinian evolution predicts the gradual emergence of new life forms in the history of life. But the fossil record tells a different story. Journey with Professor Paul K. Chien to Chengjiang, China, and the world's most extraordinary Cambrian fossil site. As he shows, this fossil site (along with many others around the world) points not to gradual evolution but to the sudden appearance of entirely new animal body plans. The best explanation? Intelligent design.

Library Cataloging Data

Biology's Big Bang: The Cambrian Explosion by Paul K. Chien

42 pages, 6 x 9 in.

ISBN-13 Paperback: 978-1-63712-061-3, Kindle: 978-1-63712-062-0, EPub: 978-1-63712-063-7

BISAC: SCI054000 SCIENCE / Paleontology

BISAC: SCI027000 SCIENCE / Life Sciences / Evolution

BISAC: SCI031000 SCIENCE / Earth Sciences / Geology

Publisher Information

Discovery Institute Press, 208 Columbia Street, Seattle, WA 98104

Internet: https://discovery.press/

Published in the United States of America on acid-free paper.

First Edition, First Printing, August 2024.

CONTENTS

BIOLOGY'S BIG BANG: THE CAMBRIAN EXPLOSION

Paul K. Chien

BY NOW MANY PEOPLE HAVE HEARD OF THE TERMS "ANIMAL BIG Bang" or "Cambrian Explosion." These terms refer to the sudden, virtually simultaneous appearance of most of the major animal groups in the early part of the Cambrian period about 530 million years ago. Many people's first exposure to the concept came from the cover of *Time* magazine in December 1995. Boldly printed across the cover were the words "Evolution's Big Bang." The subtitle read, "New discoveries show that life as we know it began in an amazing biological frenzy that changed the planet almost overnight."

Evolution, according to mainstream theory, should be a slow, gradual process involving millions of accidental mutations and intermediate steps over a myriad of generations. How could terms such as "big bang," "explosion" and "amazing biological frenzy" describe such a plodding process? For many, both inside and outside the community of evolutionary scientists, it has sounded like a contradiction in terms.

Shortly after I saw the *Time* magazine cover story, a friend alerted me to two related articles in an official Chinese paper, *People's Daily*. One was entitled "Chengjiang Fossils Challenge Evolution" and used the term "Cambrian Explosion of Life." It was about an exciting Cambrian fossil find in Chengjiang County in Yunnan Province, located in southwestern China. The marine fossils there are so astonishingly well preserved that the location has been designated a World Heritage Site by the United Nations Educational, Scientific and Cultural Organization

(UNESCO).[1] The other article concluded that further study of these extraordinary marine fossils could prove a serious blow to traditional Darwinian theory.

I myself am Chinese, so now I was doubly intrigued. There was also this: the Cambrian fossils were marine fossils, and I am a professor of marine biology, so it's not hard to imagine how eager I was to study these Chengjiang fossils, firsthand if possible.

Body Plan Bonanza

ONE REMARKABLE thing about the Cambrian explosion generally, and the Chengjiang fossil find specifically, is the extraordinary diversity of animal forms. To convey what I mean exactly, I need to unpack the concept of body plans.

I taught marine biology at the University of San Francisco for some forty years, where I remain an emeritus professor; and one of my students' favorite off-campus activities was exploring tide pools along the rocky shores of the Pacific Ocean. At low tide, a multitude of intertidal animals were exposed, allowing my students to learn firsthand where the animals lived, what they ate, how they are adapted to their particular environment, how they reproduced, what their role was in the community, and much more. But before learning all that, my students were required to identify the animals by their scientific names: genus and species. For example, the scientific name of the most common California hermit crab is *Pagurus samuelis*. For some students, identifying the animals was a difficult task. Many closely related species look alike, especially in field conditions without the help of microscopes and without being able to dissect the specimens or use a proper reference key. But most students had little difficulty determining which major group each animal belonged to, even at first glance.

Why? Because these major groupings were based on highly distinct body plans, and the body plans were dramatically distinct one from another. Even when the students found a new animal they had never seen before, they could easily tell to which major group it belonged.

In scientific terms, these distinct animal body plan groupings are known as phyla. The division between phyla is highly distinct, and there are few, if any, intermediates between phyla. For example, the clams and mussels belong to the phylum Mollusca; the crabs and shrimp are grouped into the phylum Arthropoda; and most of the worms my students and I saw were in the phylum Annelida. To give you a sense of how broad and basic a phylum category can be, all mammals belong to the phylum Chordata, as do fish, amphibians, reptiles, birds, sea squirts, and lancelets. There is tremendous variety in there, but there are body plans so distinct from these body plans that they belong in a different phylum from Chordata, while the Chordata, for all their astonishing diversity, belong in the same phylum, due to the fact that they share certain very basic architectural features.

There are dozens of phyla, and according to standard evolutionary theory, these dramatically distinct body plans arose through a series of gradual changes over numerous generations, initially with only one ancestral species of a single phylum, slowly diversifying into two and then more. This process is understood as gradual and slow, one small mutational step at a time.

According to traditional evolutionary theory, all of biology started from an ill-defined single form called the last universal common ancestor, which evolved into two, then into more and more through time in a branching tree pattern as shown in many textbooks and museums. According to this model, the new forms develop from the bottom up, so to speak: one species evolves into two species... then a new genus... then a new family... and eventually there is enough difference among some of the forms that it makes sense to identify distinct phyla. Small differences followed by bigger differences followed by differences so large that there now exist forms with completely distinct body plans.

However, the pattern of the fossil record reported in *Time* magazine, the *People's Daily*, and later by others showed a pattern completely different from that of a slowly branching tree. According to present estimates, twenty of the thirty-three still living metazoan phyla, including seventeen of the twenty-seven living bilaterian animal phyla, appeared relatively suddenly in many places around the world during the Cambrian period—thus the labels "Cambrian Explosion" and "Animal Big Bang."

People's Daily mentioned that the rich fossil site in Chengjiang County was readily accessible from the city of Kunming in southwest China. Immediately I thought to myself how wonderful it would be to visit the fossil site someday and find out all about this enigmatic event. If I could, I would share the facts with my students and friends.

It was purely a passing thought. I doubted the Chinese government would grant an American access to the site, a crown jewel among the world's fossil sites, so I never believed for a minute that this dream could come true—and so soon.

Heading to the Field

Less than a month later, out of the blue, I was surprised and privileged to be asked to organize an international team of scholars to visit the fossil site and meet with the paleontologists who had made the big discovery. It turned out that these fossil experts were based at the Chinese Academy of Sciences in Nanjing, in the eastern part of the country and far from the fossil site. For me to organize such a complex group trip to Nanjing in eastern China and Chengjiang in western China proved a daunting task, even though I had several years of experience traveling in China on my own to teach biology in summer schools. International academic exchanges with China at this scale were not a common occurrence. However, with the generous help of many people and, later, the full cooperation of the Chinese scholars and the government, we made all the connections, solved all the financial issues, obtained all the neces-

Figure 1. Author Paul Chien in front of Maotian Shan (Mt. of Heavenly Hat) near Chengjiang, China, during a later visit.

sary approvals, and waded through all the red tape, mostly by snail mail. Five months after the invitation, our team arrived.

Our group included Professor James Valentine from the University of California at Berkeley, and W. Y. Leung, chair of the Communication Department at the Chinese University of Hong Kong. Professor Leung also convinced a famous television director from Hong Kong to bring a small film crew along to record the visit. It was the first television team outside of China to film the Chengjiang site. At that time, bringing in specialized camera equipment and filming in China required very special permission. Looking back, it seemed like a miracle, and I still wonder how so many closed doors were opened for us one after another, and in such a timely fashion.

First stop on our trip was Nanjing. At the Chinese Academy of Sciences there, we received a very warm welcome from the director, and met many scientists working in related fields. We also visited the labs of two principal investigators, where we got our first glimpse of the oldest, and extremely well-preserved, marine animal fossils of many different phyla. These creatures already exhibit bilateral symmetry, differentiated appendages and digestive tracts, and well-developed brains, as well as eyes. Valentine noted how the phyla and classes were fully developed when they first appeared in the Cambrian.

During this visit, the Chinese Academy of Sciences also arranged a full-day symposium about the fossils. In the morning, we heard talks from major researchers in the field, followed by a question-and-answer period. In the afternoon we were given a lot of time to exchange information and ideas about the explosion. Most interesting to me was the discussion on the possible causes of the sudden concurrent appearance of so many different animal body plans. Everyone seemed to agree that the explosive appearance of so many phyla without apparent ancestors contradicted the classic Darwinian model, both its original nineteenth-century form and the updated neo-Darwinian model developed in the wake of the twentieth-century revolution in genetics and molecular biology. The scientists and scholars realized that random genetic mutations acted upon by natural selection were not capable of generating so many disparate groups of animals in the short time available.

There was some disagreement on how long the Cambrian explosion took. The commonly cited time period in the literature is 20-30 million years, but some of the Chinese researchers seemed to think that the main thrust of the explosion took only 1-3 million years. The yellow shale layers containing most of the phyla were not very thick at all. Still others tried to explain away the seeming suddenness of the event. In any case, even assuming the wider window of time, and even though this seems like a long time to us, from the perspective of geology it is quite sudden for producing so many new phyla. The traditional Darwinian story simply did not fit with the fossil record, and new ideas and explanations were needed.

One of the Chinese scholars who spoke offered the common, still debated, idea that a sudden increase in the oxygen level in the ocean caused the explosion. Others theorized that perhaps the accumulation of nutrients in the ocean at that time, or shortly before, might have triggered bacterial and algal blooms that would in turn be a food source to feed a rapid development of animals. But while interesting, these explanations focus on *necessary* conditions for the Cambrian explosion (oxygen and a nutrient source), but do not offer a *sufficient* cause for the sudden emergence of all these animal body plans. It would be like claiming that because birds fly in the air, the existence of the Earth's atmosphere somehow caused birds to appear. Such explanations just don't make sense logically or practically.

A visitor to the meeting suggested another explanatory avenue: we could start comparing Hox genes in different animal groups. Similar Hox genes exist in many different species, and they help control the development of embryos. Slight changes in the Hox genes might have quickly given rise to different body plans, it was suggested. This was a new idea to many paleontologists at that time and seemed interesting, but where did the Hox genes come from in the first place? Later we learned that Hox genes often function as on/off switches of other coding genes or specify the positioning of various biological structures, but do not themselves transmit all the required biological information about body plans.

One scholar suggested that evolution might be a combination of chance and necessity working together, in that once a new body plan got established by chance, then by necessity it would radiate into different forms within the same body plan in the new environment. For example, when arthropods first showed up in the Cambrian period, a large number and variety of slightly different forms of arthropods immediately followed. However, arthropods seemed to be the only example; other phyla did not follow this pattern. Also, at best it would explain the rapid emergence of diversity within a phylum, not the sudden origin of so many distinct phyla in the first place.

The punctuated equilibrium model proffered by Stephen J. Gould and Niles Eldredge was brought up. On this model, species generally remain stable, showing little or no change over time in the fossil record, but when change occurs, it occurs rapidly in rare and isolated locations, leaving little record of the change in the fossil record. Punctuated equilibrium was based on the observation that species appear suddenly in the fossil record, but it does not provide an explanation of how the many body plans of the Cambrian could actually have arisen in the time available.

Thus it is that Sean B. Carroll, a mainstream evolutionary developmental biologist firmly committed to modern evolutionary theory, can state confidently, "The explosion of animal diversity in the Cambrian is one of the most important and compelling mysteries in the history of life." And he writes this in an endorsement of a relatively recent book on the Cambrian explosion widely regarded as a benchmark in the field and one that also emphasizes the persistent mystery of the dramatic transition from Precambrian sponges to the world of the Cambrian with its riot of animal body plans.

In that book, *The Cambrian Explosion*, authors Douglas Erwin and James Valentine remain committed to searching out a purely material evolutionary account of the Cambrian explosion, and yet they insist that this singular event in the history of life remains marked by important "unresolved questions," and they call the transition from sponges to the creatures of the Cambrian "the most enigmatic of any evolutionary transition in metazoans."[2]

The point is central enough that Christopher J. Lowe highlights it in his review of the book in the journal *Science*. "The grand puzzle of the Cambrian explosion surely must rank as one of the most important outstanding mysteries in evolutionary biology," he writes.[3]

Over the course of the symposium there was plenty of speculation, but no answers. The central question is what produced all those new body plans in a short period of time. No one seemed to have a good answer, and it felt like there was little progress. But I wasn't discouraged, because it struck me that here in a thoroughly academic setting the Cambrian event was being openly recognized as a unique "explosion" and a serious challenge to modern Darwinian theory, which was a big step forward for intellectual freedom of inquiry. Then, too, I knew that studies in Chengjiang had just begun; there was much to be learned and more data yet to be collected.

Chengjiang's Exquisite Cambrian Fossils

THE DISCUSSION at the symposium in Nanjing was fascinating, but for me it was only a prelude to what I most eagerly anticipated: our visit to the Chengjiang fossil site itself, arguably the best Cambrian fossil site in all the world. It did not disappoint.

Shortly after the symposium, we set out for the Chengjiang site with Chinese paleontologist J. Y. Chen and his colleagues, including Ms. Zhou, a highly respected paleontologist from the Nanjing Institute of Paleontology, who had a fossil, *Misszhouia*, named after her. Our journey began with a plane ride more than 1,000 miles across southern China to the southwestern city of Kunming. After departing from the plane, our group squeezed into a small car with all our gear, and after a nearly two-hour ride past villages and over the mountains, we settled into a quaint hotel nestled in the agriculture town of Chengjiang in the middle of tobacco fields. As we later explored our surroundings, I realized this hotel was the fanciest one in town.

Early the next morning, we took a short ride up the red rolling hills behind the town. The bouncing dirt road led us past a smoking phosphate mine processing facility and ended near a small quarry of loose yellow rocks, quite unremarkable in appearance. It looked just like several other hillsides in the region. Little did I know that this was the original site of one of the greatest discoveries of animal fossils—a place most paleontologists all over the world would love to visit.

We piled out of the vehicle and traipsed the short distance to the unassuming quarry. Ms. Zhou knew this place like the back of her hand. She skillfully wielded her geological axe a few times and cracked open some ordinary-looking yellow rocks. Suddenly an exquisitely preserved shrimp-like animal appeared in front of our eyes! Although the specimen was over 500 million years old, we could clearly see the fossilized image of its eyes, antennae, legs, and even the hairs on its legs. No wonder paleontologist Xian-Guang Hou once remarked that one of the Chengjiang fossils looked "as if it was alive on the wet surface of the mudstone."[4] I eagerly followed Ms. Zhou's lead and found half a dozen fossils myself in an hour. It seemed that everywhere we looked at the site we came across more of these exquisite, ancient fossils.

I want to share with you some photos of these remarkable fossils my colleagues and I have taken over the years. Some were taken on my first visit to Chengjiang.

One particular fossil fish, *Myllokunmingia fengjiaoa*, was discovered by Professor D. G. Shu and reported on in the journal *Nature*.[5] I was privileged to visit Dr. Shu's lab, and was thrilled when he took out this specimen and I was able to examine it in person under a light microscope.

The Chengjiang site is not the only Cambrian site to testify to the remarkable nature of the Cambrian explosion. For instance, the Burgess Shale in Canada (more on that remarkable fossil site below) also points to the abrupt appearance of many new phyla during the Cambrian. One of the more remarkable fossils from this Canadian site is *Metaspriggina*, a 500-million-year-old fish fossil (Lower–Middle Cambrian) showing a pair of well-developed eyes. There is little doubt that both invertebrate compound eyes and vertebrate camera eyes were already well developed early in the Cambrian era.[6]

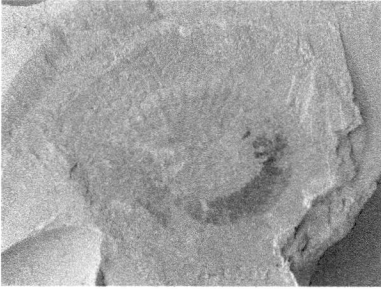

Figure 2. *Stellostomites*, a disc-shaped, soft-bodied animal with a well-developed u-shaped gastrointestinal tract (the dark curved section at the lower right of center). Modern jellies belong to a different phylum and do not have this gastrointestinal tract.

Figure 3. Triangular structures surrounding the *Stellostomites* are conical shells of Hyoliths. Recent studies consider both as filter feeders related to other phyla, such as Phoronida and Brachiopoda, not sea jellies and mollusks.

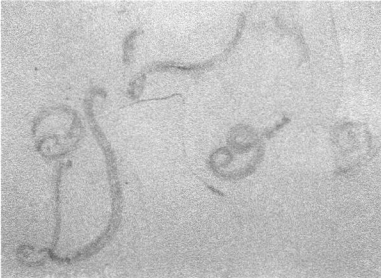

Figure 4. Worms from the species *Maotianshania cylindrica*.

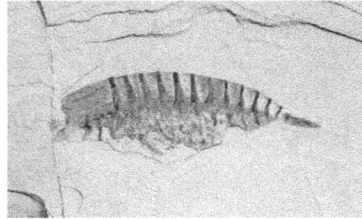

Figure 5. The arthropod *Leanchoilia* is found in both the Chengjiang deposits, as well as the Burgess Shale in Canada.

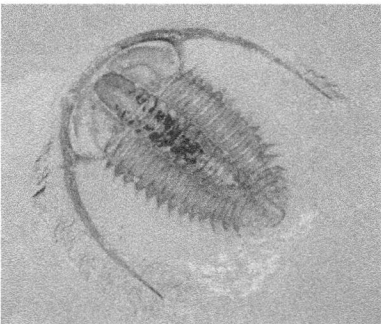

Figure 6. A well-preserved trilobite specimen from the Maotianshan Shale.

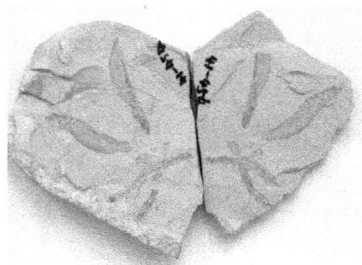

Figure 7. Hundreds of *Haikouella*, phylum Chordata, were found near the Chengjiang site in 1999.

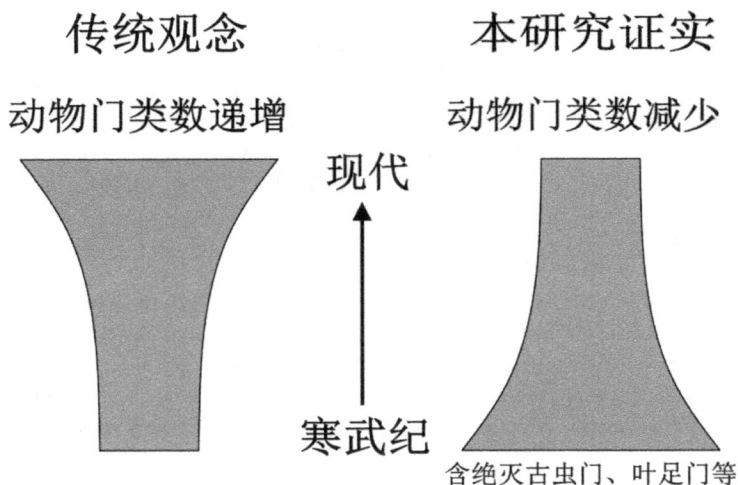

传统观念　　　　本研究证实

动物门类数递增　　　动物门类数减少

现代

寒武纪

含绝灭古虫门、叶足门等

Figure 8. Recreation of a graphic provided by Professor Shu to the author, comparing the expected increase of phyla over time under the traditional evolutionary model (*left*) versus the actual data showing a large number of phyla originating in the Cambrian and a loss of some of those phyla over time (*right*).

Top Down vs. Bottom Up

ON ANOTHER trip, a few years after my initial trip, I took a television production team from Hong Kong to visit Professor D. G. Shu and his lab at North West University in Xian, China. He gave us his diagram comparing the traditional concept of the development of animal phyla with his own study and conclusion. The traditional model predicted that the number of animal phyla would gradually increase with time (left side drawing in Figure 5.8), beginning with a single or a few phyla. However, his research turned the traditional model upside down, showing that at the beginning of the Cambrian, most of the animal phyla appeared abruptly, and the number of Cambrian phyla decreased over time, by extinction (right side drawing). This pattern in the Cambrian fossil record may be as damaging to Darwinian theory as the sudden appearance of phyla.

Avoiding the Hard Facts

IN THE 1990s, before the Cambrian explosion data from the Chengjiang fossil site was widely known, a museum in Golden Gate Park, San
Francisco, featured an exhibit called the "Hard Facts Wall." On the wall,
hard rocks containing fossil specimens were displayed and arranged in
the pattern of a tree, as if the hard facts supported the idea that the history of life had followed a branching-tree pattern as predicted by Darwin.

There was just one problem. It wasn't true.

A trained geologist, John Wiester, was suspicious. Wiester, a member of the American Scientific Affiliation and its Committee for Integ-
rity in Science Education, looked into the ages of the fossils on display
and found that many of the fossils were not placed in the proper geological layers according to their age. Some of the older fossils were placed at
the same level as younger fossils, while some of the younger fossils were
placed in older geological layers. The overall history of these fossils could
be fit into the pattern of Darwin's tree only by twisting the data.

Appalled, Wiester wrote an article entitled "Shell Games in California" showing that if the fossils had been arranged according to their
true ages, they would show a pattern of parallel straight lines, with each
line representing an animal phylum and the base of these isolated lines
appearing abruptly at about the same time, around 550 million years
ago.[7] In colloquial terms, this would be a "lawn model" rather than a
tree model.

Wiester's analysis of the museum fossils gained corroborating support from what paleontologists were uncovering at Chengjiang. The data
from this fossil site, released since 1995, substantiates this concept of a
lawn model rather than a tree model. All the animal phyla there appear
abruptly in the geological column, close together in time, and with no
clear connections between them.

When I later learned of the Hard Facts Wall exhibit, I wondered
why the museum staff would create an exhibit that was inconsistent with

18

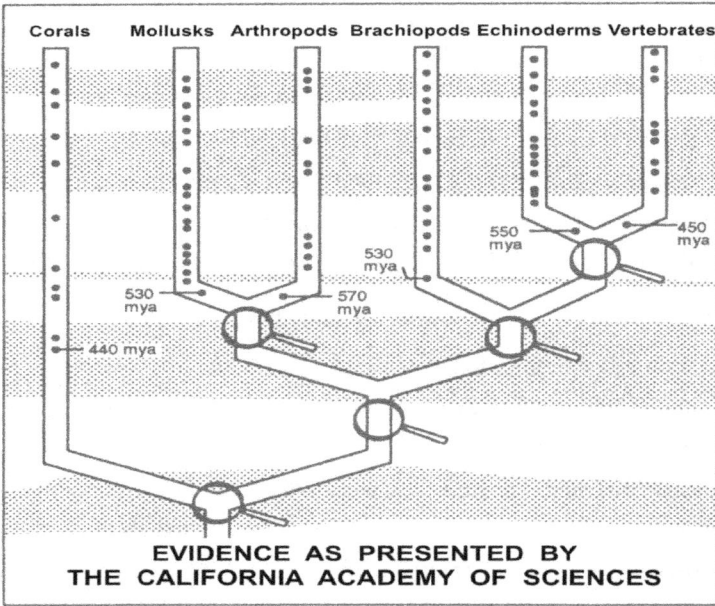

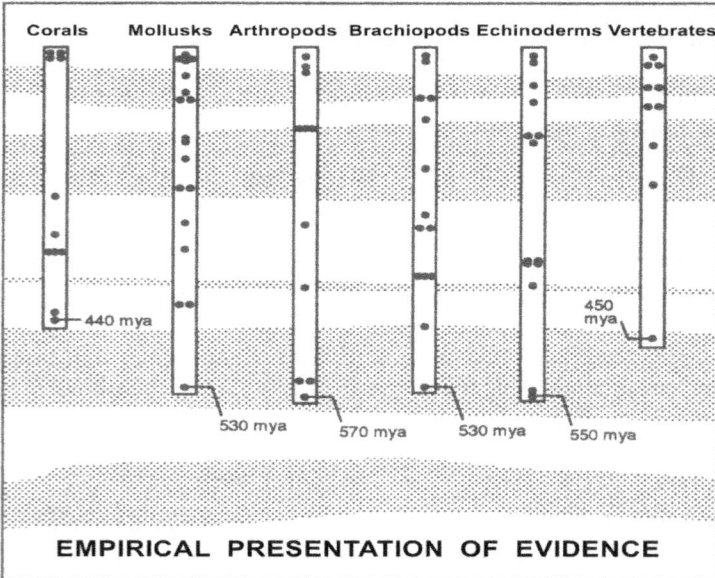

Figure 9. Representation of the Hard Facts Wall exhibit (*top*),
and of the actual fossil data (*bottom*). Notice on the bottom the
abrupt appearance and the lack of a branching tree pattern.

the actual data. The museum staff could perhaps be forgiven for not knowing about the Cambrian fossil discoveries in China that were starting to shake up the world of paleontology,[8] but why, I wondered, would the exhibit purposely misrepresent the ages of the various fossils just to make it look as if Darwin's prediction of a branching tree were true?

Another interesting feature on the original Hard Facts Wall was a series of large magnifying glasses placed over every branching point of the evolutionary tree. But, ironically, if viewers looked closely, under every magnifying glass, without exception, there were no fossils, just empty space. In other words, according to Darwin's theory, at each branching point there was supposed to be a common ancestor, but the museum did not, and could not, show any fossils of a common ancestor between two groups, and for a very simple reason: no such common ancestors had been found.

If museum visitors didn't look closely, the Hard Facts Wall looked like an impressive confirmation of Darwin's prediction. But to the trained eye, the Hard Facts Wall inadvertently demonstrated that the pattern of the fossil record contradicts that prediction.

Thankfully, this Hard Facts Wall exhibit did not survive when the museum reopened after a major renovation. On a new wall there was a display entitled "Timeline of Life on Earth," showing major events starting 4.6 billion years ago when the Earth was formed and continuing to the later start of life on Earth and onwards.

In two separate visits soon after the renovation, I looked to see how they would show the Cambrian explosion, which by that time was much better known in the scientific literature. Unfortunately, it turned out that the most important discovery in paleontology of the twentieth century was completely missing. The major events shown on the museum wall jumped from 650 million to 450 million years ago, completely skipping the Cambrian explosion! Again, I wondered why the museum would omit this critical piece of evidence that challenged Darwin. I still wonder if this evidence will eventually be included.

Figure 10. California Academy of Sciences museum display of the "Timeline of Life on Earth," missing one of the most important discoveries of the twentieth century in biology: the Cambrian Explosion, approximately 530 million years ago.

Beijing National Museum of Natural History

I HAD another, more positive, museum experience halfway around the world. Several years after my first visit to Chengjiang, I visited the National Museum of Natural History in Beijing. To my pleasant surprise, I found a large room housing an exhibit called the "Cambrian Big Bang of Life." A beautiful collection of Chengjiang fossils were on display. In the summary diagram, the museum showed a lawn model to illustrate the development of animal phyla since the early Cambrian period, instead of the textbook standard Darwinian tree model. The Beijing diagram was much more consistent with the actual fossil record.

In the Beijing diagram, vertical solid parallel yellow lines were used to represent the history of animal phyla. The phyla were correctly depicted with most of them beginning in the early Cambrian, and the parallel lines with no interconnections between them showed that there was no known evolutionary relationship between the phyla since that

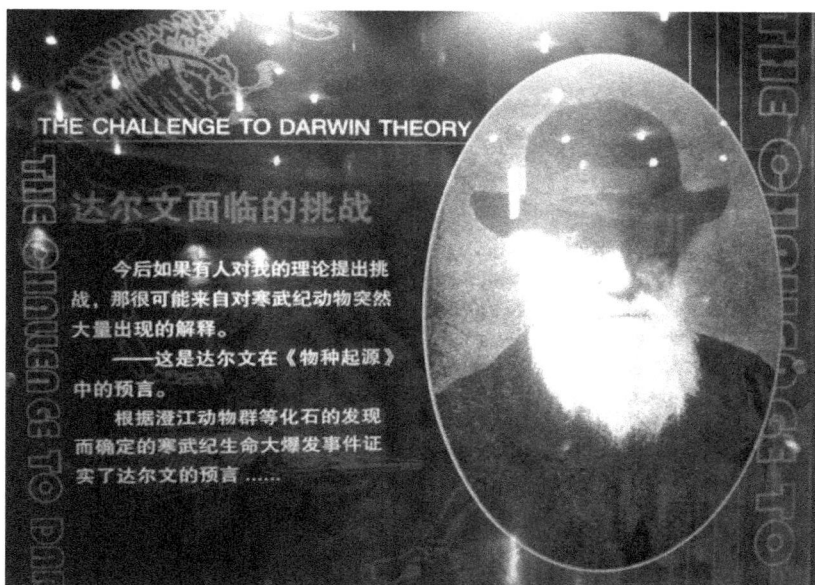

Figure 11. Panel from a fossil exhibit at the Beijing National Museum of Natural History, noting the challenge to Darwin's theory posed by the Cambrian fossils at Chengjiang.

time. Only the solid lines for the sponges and mollusks extended earlier into the late pre-Cambrian (reflecting that sponge and mollusk fossils had been found predating the other phyla), and the other phyla had only dotted lines before the Cambrian, raising the question of whether they had an earlier origin. It was gratifying to see that the Beijing exhibit was much more accurate than the San Francisco museum's exhibit, showing the actual data rather than trying to uphold the Darwinian tree model in the face of contrary evidence.

Perhaps more remarkable was the concluding panel of the Beijing exhibit, quoting a concern Darwin himself had raised in *The Origin of Species*. Darwin acknowledged that the sudden appearance of a large number of animals during the Cambrian period might challenge his theory. The last panel of this museum display concluded by observing that the discovery of the Chengjiang fauna and other fossils had confirmed the reality of the Cambrian Explosion, confirming Darwin's concern and underscoring the challenge to his theory.

I was grateful for the remarkable opportunity I had received several years earlier to visit Chengjiang and see firsthand in the field the fossils that challenge Darwin's theory. The Beijing exhibit gave me hope that perhaps more people would start to learn the truth about the fossil record.

Reflecting on these two very different museum experiences, I am reminded of the ironic observation made by Chinese paleontologist J. Y. Chen when answering questions after a lecture he gave in the United States on one of his visits: "In China, we can criticize Darwin, but not the government. In America, you can criticize the government, but not Darwin."[9]

A Mountain of Evidence

A FEW years later, in the early 2000s, I was privileged to participate in another discussion about the cause of the Cambrian explosion, this time on a mountaintop in southwestern Canada.

Several Canadian geologists, along with researchers from Discovery Institute's Center for Science and Culture, invited me to join them on a special guided hike up Wapta Mountain in Canada's Yoho National Park to visit the Burgess Shale, the most famous Cambrian fossil site in the western world. This site was discovered in 1909 by Charles Walcott, secretary of the Smithsonian Institution of the United States. This was the site all Chinese paleontologists I knew would consider an honor to visit. To me, the opportunity was the chance of a lifetime, and I happily accepted the invitation without hesitation.

Our group met in a small parking lot at the base of the mountain on a sunny day in July. It took us more than four hours to make the hike. We needed to stop and rest often, especially on the last stretch. At some points the trail seemed to go straight up (not actually, but it felt like it!) and the air was thin.

When I finally reached the site of the Burgess Shale, I looked around at the surrounding snow-covered mountains, glaciers across the valley, and the quarry right beside me, trying to imagine how Walcott and others worked there a century ago.

For visitors now, it is prohibited to collect fossils or even any rock samples at the site, but we could turn over loose rocks left by previous workers. I found a variety of partly broken marine invertebrate fossils scattered around. I could recognize many well-preserved arthropods, worms, sea jellies, and brachiopods. Most of them were similar to those I had seen in China, but there were a few unique fossil species I had not seen before. However, they all belonged to the same set of body plans present in both locations.

Our guide, who held two PhDs in related fields, gave an excellent introduction to the history of discovery and studies since Walcott. He opened a steel lockbox kept at the site, brought out a wonderful collection of exquisite fossils representing many phyla, and gave an interesting talk on each of them. In his concluding remarks, he praised the wonderful process of evolution for producing such a rich treasure millions of years ago and stated that without evolution there would be no humans now.

The youngest member of our group, a teenager, asked our official guide a very simple question: "Where did all the new DNA come from?"—meaning that an explosive appearance of very different body plans and different kinds of organisms must need a significant amount of new DNA to code for them. Where did such diverse sets of new DNA come from in this explosive event?

Our guide had seemingly never thought of this question before, as he hesitated for a few seconds and finally acknowledged, "That is a very good question." It seemed to me that he had decided that evolution must somehow have produced all the new DNA, and that he had decided this without even considering how this could have occurred. Hopefully the honest question of an inquisitive teenager gave our guide something to think about in the days after our visit.

Cambrian Quagmire

HAVING REFLECTED on the many discussions about the potential causes of the Cambrian explosion, both in the scientific literature and in conversations that I have participated in stretching back to my initial trip overseas to visit to the Chengjiang fossil site twenty years ago, it strikes me that not much progress has been made along traditional lines. Most people are still bogged down in some form of the Darwinian framework even though the fossil record suggests something quite different.

Some do try to think outside the traditional box. In November of 2016, many distinguished biologists and scientists met at a conference hosted by the Royal Society of London, one of the most distinguished scientific organizations in the world. A key area under discussion was the growing dissatisfaction with the neo-Darwinian explanation for the generation of biological novelty.

Two years later, an announcement for a conference held in Salzburg, Austria, was even more direct in its critique of neo-Darwinism: "For more than half a century it has been accepted that new genetic information is mostly derived from random, error-based events," the announcement read. "Now it is recognized that errors cannot explain genetic novelty and complexity."[10]

What sort of progress have evolutionists made in coming up with a purely materialistic alternative to modern Darwinism for the Cambrian explosion? The situation has become so desperate that recently several scientists from different fields joined together to propose that the Cambrian animals, as well as the first life on Earth, came from outer space.[11] There is precious little evidence for this extra-terrestrial theory. I see this kind of proposal as an admission of how existing evolutionary explanations for the Cambrian explosion, Darwinian and otherwise, have failed.

In contrast to these proposals, which try to provide a purely naturalistic explanation for the explosion of life in the Cambrian, philosopher of science Stephen Meyer and others have proposed intelligent design as the best explanation. The idea certainly fits the fossil data much better than does traditional evolutionary models.[12]

My Study of the Precambrian Sponge Embryos

FINALLY, I want to share with you my own experience studying some of the remarkable early life forms on Earth, and briefly explore the light it sheds on the question of animal origins.

During my studies as a graduate student in the 1960s, I learned techniques to study the structure of living animal tissues and cells using electron microscopes. Mastering these techniques turned out to be very useful when applied to minute specimens in rocks.

Years later, as I was cooperating with scholars from different parts of China, we found large numbers of nearly perfectly spherical objects in 570-million-year-old phosphorus-rich rocks from Guizhou province, China, just east of Kunming City. This is from the period just before the Cambrian layers. The rock samples were cut into half-inch slices and glued onto microscopic glass slides. Then the rock slices on the glass were carefully ground by hand down to wafer-thin so that we could study them through light microscopes.

In many of the thin rock slides, we found many microscopic round fossilized objects. Some of these spheres were algal cell fossils. These were easy to identify by their thick cell walls and daughter cells that tend to adhere together, sharing a common cell wall as they divide. But a large percentage of the spheres appeared to be the fossils of sponge cells and embryos with characteristic spicules. (No other known group of animals contains spicules.) The sponge eggs and early embryos were in the range of 0.6 to 0.7 mm in diameter. In 1999, we presented our findings at a scientific conference in Kunming, China, sponsored by the Early Life Research Center and the Chinese Academy of Sciences.[13]

My techniques in scanning electron microscopy came in handy in later research. In those further studies, I photographed these sponge eggs and early embryos at much higher resolutions. After carefully cracking them open and using a scanning electron microscope, I could identify cellular structures inside the cells, such as the nuclei and granules of egg yolk, that light microscopy could not resolve. In 2001, my colleagues and I presented another paper illustrated with scanning electron micrographs at a conference at the University of California, Berkeley, detailing our discoveries.[14]

During this work I was amazed by the discoveries we made, but as I look back now after more than a decade, I find that I am even more amazed by what we didn't see. When my colleagues and I searched through thousands of 570-million-year-old thin slides of Precambrian rock samples and photographed thousands of isolated objects under a scanning electron microscope, we found only sponges and algae, and no forms of life that even approached what one might deem bilaterian animals. The humble sponges were as close as it came, and even the identification of Doushantou/Weng'an fossils as sponge embryos has recently been disputed.[15]

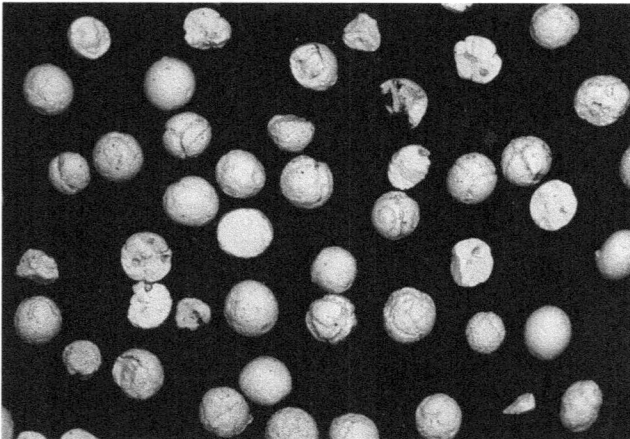

Figure 12. Electron microscope image of 570-million-year-old small round fossils of sponge specimens and algae.

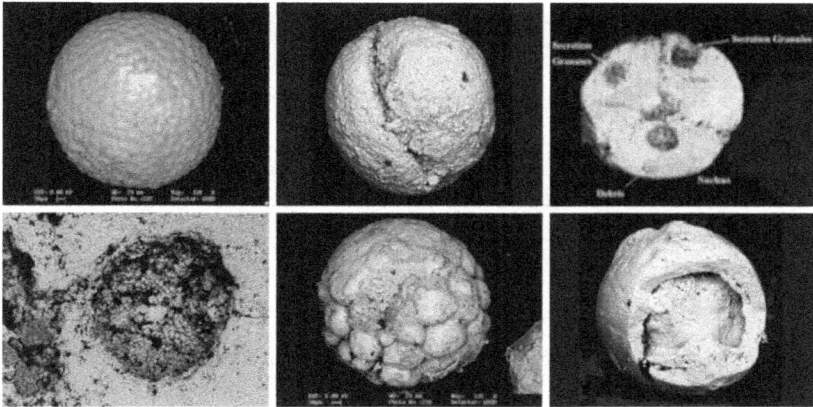

Figure 13. *Top left*: Well-preserved fossil of sponge egg cell with the outer membrane intact. *Top center*: Fossil of sponge embryo at the two-cell stage with the outer membranes removed. *Top right*: Fossil of sponge embryo cracked open showing cellular content of three cells. *Bottom left*: Enlarged and rotated image of the same fossilized embryo, showing the round nucleus in three dimensions at the center of the cell. *Bottom center*: Embryo stage where more than thirty cells can be counted. *Bottom right*: Fossilized sponge embryo at a more advanced stage of development.

My understanding is that scientists and their students at several labs in China have also studied those Precambrian rocks and confirmed our discoveries of sponge eggs and embryos. Adult sponge body fossils were also reported. Some evolutionists were hoping to find more animal remains that could be claimed as Cambrian precursors. So far, the rocks refuse to yield the evidence these researchers were hoping for. The problem is exacerbated by the fact that evolutionary theory predicts countless missing links between sponges (or some still earlier, simpler life form) and the Cambrian animal phyla. The theory needs countless transitionals, but continues to go begging for even a very few.

Some have tried to rescue evolutionary theory by claiming that there might have been many precursor animals leading up to the Cambrian, and it's just that Precambrian conditions were not very good at preserving those fossils, so those precursors are missing from the fossil record. But if the conditions for fossil preservation were so poor, why did they manage to preserve soft, delicate sponge eggs and early embryos, and preserve them extremely well, including the nucleus in eggs and embryo cells? Given this, why have no precursors to the Cambrian animals yet been found?

Alternately, if—as a few Precambrian fossil specialists have suggested—the Doushantou/Weng'an fossils are not sponge embryos, then one might again try to argue that the Cambrian explosion is merely an artifact of an incomplete fossil record. But any attempt to dismiss the Cambrian explosion as mere illusion, with or without Precambrian sponge embryos, flies in the face of mounting evidence.

As German paleontologist Günter Bechly notes,[16] vast fossil troves of Ediacaran-age fossils have recently been discovered in Mongolia and China,[17] and these sites lack any bilaterian animals and have only yielded fossil algae. The fact that these rocks preserve soft-bodied fossils like algae is significant because these localities are of the Burgess Shale type, which shows that the preservation conditions were capable of preserving small, soft-bodied organisms—exactly the type of creatures that are proposed to be the ancestors of Cambrian animals. The fact that they do not preserve anything that resembles such animal ancestors indicates that those animals were simply not present. Even a recent paper in *PNAS* that tried to downplay the Cambrian explosion acknowledged that these new sites show that animals are unknown from the Ediacaran not because of preservational issues, but because they definitely did not yet exist.[18]

What of those who interpret certain trace fossils as suggesting possible animal forms in the Precambrian, thus partially mitigating the astonishing efflorescence of new animal body plans in the Cambrian? "The Ediacaran record falls far short of establishing the existence of the wide variety of transitional intermediates that a Darwinian view of life's history requires," Meyer comments. "The Cambrian explosion attests to the first appearance of organisms representing at least twenty phyla and many more subphyla and classes, each manifesting distinctive body plans. In a best case, the Ediacaran forms represent possible ancestors for, at most, four distinct Cambrian body plans, even counting those documented only by trace fossils. This leaves the vast majority of the Cambrian phyla with no apparent ancestors in the Precambrian rocks."[19] Further, the majority of alleged animal trace fossils from the Ediacaran has been refuted by a recent experimental study, which exactly reproduced all these traces as artifacts of stirred up bacterial mats.[20]

And as we saw at the beginning of this mini-book, it isn't just Meyer and other design theorists. Indeed, that the Cambrian explosion was a real event is the mainstream view of Cambrian paleontologists. As Erwin and Valentine emphasize, "Several lines of evidence are consistent with the reality of the Cambrian explosion."[21] Or as observed by Martin Scheffer, a Dutch ecologist, winner of the Spinoza Prize, and a member of the US National Academy of Sciences, "It could be that earlier rocks were not as good for preserving fossils," but we now know that "well preserved fossils do exist from earlier periods, and it is now generally accepted that the Cambrian explosion was real."[22]

Listening to the Whispers of the Past

Darwin recognized that the fossil record posed a serious difficulty for his theory. He hoped that future discoveries would overturn the picture and confirm his prediction of a slow, gradual, step-by-step evolutionary process, complete with a fossil record that looked like a branching tree. However, since Darwin's time the fossil record has stubbornly refused to confirm his prediction. Instead, as we have discovered more—including the remarkable fossils sites in China and Canada testifying to the astonishing diversity and suddenness of the Cambrian explosion—matters have only gotten worse for Darwin's story.

What are we to do with these findings? Rather than pretending Darwin's tree is still healthy, rather than hiding data or presenting a one-sided picture for museum attendees, we need to find the courage to accept the fossil record for what it is, including the fascinating record of the abrupt appearance of numerous animal phyla during the Cambrian in an explosive burst of creativity. Then we need to follow this evidence, this insistent pattern, to the best explanation. I submit to you that the best explanation invokes the only cause with the demonstrated ability to generate new biological form and information so quickly. It is the one cause known to create in a top-down pattern such as we find in the fossil record. That cause is intelligence.

Review: Your Turn

1. Where did Paul Chien go to see fossils from the Cambrian period?

2. Why do scientists refer to a "Cambrian Explosion?" What is it about the appearance of these animals on the Earth that is like an "explosion" or a "Big Bang"?

3. What is it about the Cambrian Explosion that challenges Darwin's theory of evolution?

4. Why do you think the San Francisco museum exhibit placed the fossils in the shape of a tree, with some fossils in the wrong place on the timeline?

5. How was the Beijing museum's exhibit different from the San Francisco museum's exhibit?

6. Why is it significant that so many soft-bodied sponge and sponge embryo fossils have been preserved right before the time of the Cambrian Explosion?

FUEL YOUR CURIOSITY!

Recommended Resources for Further Exploration:

VIDEOS

Darwin's Nightmare

Whale Evolution: Good Evidence for Darwin?

Fossils: Mysterious Origins

PODCASTS

Why Intelligent Design Best Explains the Fossil Record Data
Casey Luskin

Günter Bechly on Why Seventy Years of Textbook Wisdom Was Wrong

ENDNOTES

1. See "Chengjiang Fossil Site," World Heritage List, UNESCO, accessed February 13, 2020, https://whc.unesco.org/en/list/1388/.

2. See Douglas H. Erwin and James W. Valentine, *The Cambrian Explosion: The Construction of Animal Biodiversity* (Greenwood Village, CO: Roberts and Company, 2013), 330, 324.

3. Christopher J. Lowe, "What Led to Metazoa's Big Bang?" *Science* 340, no. 6137 (2013): 1170–71, https://doi.org/10.1126/science.1237431.

4. Xian-guang Hou et al., *The Cambrian Fossils of Chengjiang, China: The Flowering of Early Animal Life* (Oxford: Blackwell, 2004), 13.

5. D. G. Shu et al., "Lower Cambrian Vertebrates from South China," *Nature* 402 (November 4, 1999): 42–46, https://doi.org/10.1038/46965.

6. Simon Conway Morris and Jean-Bernard Caron, "A Primitive Fish from the Cambrian of North America," *Nature* 512 (June 11, 2014): 419–22, https://doi.org/10.1038/nature13414.

7. John L. Wiester, "Shell Games in California," *Origins Research* 14, no. 2 (1992): 11.

8. Even before Chengjiang, other Cambrian fossil sites around the world, the Burgess Shale especially, already suggested the lawn model of an abrupt appearance of multiple phyla close together in time. But the Chengjiang fossil discoveries made that pattern all the plainer, and the fossils were so well preserved that they created an international sensation, causing the news about the Cambrian explosion to leap well beyond the specialized world of paleontology and evolutionary biology.

9. J. Y. Chen, quoted by Stephen C. Meyer in *Darwin's Doubt: The Explosive Origin of Animal Life and the Case for Intelligent Design* (New York: HarperOne, 2013), 52.

10. See the announcement for the conference on *Evolution: Genetic Novelty/Genomic Variations by RNA Networks and Viruses*, Salzburg, Austria, July 4–8, 2018, http://www.rna-networks.at/about/.

11. Edward J. Steele et al., "Cause of Cambrian Explosion—Terrestrial or Cosmic?" *Progress in Biophysics and Molecular Biology* 136 (August 2018): 3–23, https://doi.org/10.1016/j.pbiomolbio.2018.03.004. Consider also an interview critique with biologist Ann Gauger, "Octopuses from the Sky: Scientists Propose 'Aliens Seeded Life on Earth,'" July 9, 2018, in *ID the Future*, podcast, MP3 audio, 10:32, https://www.discovery.org/multimedia/?s=outer+space.

12. For Stephen Meyer's extended case on the matter, see *Darwin's Doubt* and *Debating Darwin's Doubt: A Scientific Controversy That Can No Longer Be Denied*, ed. David Klinghoffer (Seattle: Discovery Institute Press, 2015). Other key intelligent design scholars and theorists include Douglas Axe, Michael Behe, William Dembski, Guillermo Gonzalez, Phillip Johnson, Paul Nelson, Jay Richards, and Jonathan Wells. See, for example, a partial list of prominent intelligent design scholars at "Fellows,"

Discovery Institute Center for Science and Culture, www.discovery.org/id/about/fellows.

13. J. Y. Chen, C. W. Li, Paul Chien, G. Q. Zhou, and Feng Gao, "Weng'an Biota: Casting Light on the Precambrian World" (paper presentation, The Origin of Animal Body Plans and Their Fossil Records, Kunming, China, June 20–26, 1999).

14. Paul Chien, J. Y. Chen, C. W. Li, and Frederick Leung, "SEM Observation of Precambrian Sponge Embryos from Southern China, Revealing Ultrastructures Including Yolk Granules, Secretion Granules, Cytoskeleton, and Nuclei," (paper presentation, North American Paleontological Convention, University of California, Berkeley, June 26–July 1, 2001).

15. John A. Cunningham et al., "The Weng'an Biota (Doushantuo Formation): An Ediacaran Window on Soft-bodied and Multicellular Microorganisms," *Journal of the Geological Society* 174, no. 5 (2017): 793–802, https://doi.org/10.1144/jgs2016-142; David J. Bottjer et al., "Comparative Taphonomy and Phylogenetic Signal of Phosphatized Weng'an and Kuanchuanpu Biotas," *Precambrian Research* (forthcoming). Published ahead of print, August 8, 2019, https://doi.org/10.1016/j.precamres.2019.105408; Jonathan B. Antcliffe et al., "Giving the Early Fossil Record of Sponges a Squeeze," *Biological Reviews* 89, no. 4 (April 29, 2014), https://doi.org/10.1111/brv.12090.

16. Günter Bechly, "Alleged Refutation of the Cambrian Explosion Confirms Abruptness, Vindicates Meyer," Evolution News and Science Today, Discovery Institute, May 29, 2018, https://evolutionnews.org/2018/05/alleged-refutation-of-the-cambrian-explosion-confirms-abruptness-vindicates-meyer/.

17. Stephen Q. Dornbos et al., "A New Burgess Shale-Type Deposit from the Ediacaran of Western Mongolia," *Scientific Reports* 6 (2016): 23438; Xunlai Yuan et al., "An Early Ediacaran Assemblage of Macroscopic and Morphologically Differentiated Eukaryotes," *Nature* 470 (2011): 390-3.

18. Allison C. Daley et al., "Early Fossil Record of Euarthropoda and the Cambrian Explosion," *PNAS* 115, no. 21 (2018): 5323-31.

19. Meyer, *Darwin's Doubt*, 85-6.

20. Giulio Mariotti et al., "Microbial Origin of Early Animal Trace Fossils," *Journal of Sedimentary Research*, 86 (2016): 287–93.

21. Erwin and Valentine, *The Cambrian Explosion*, 6.

22. Martin Scheffer, *Critical Transitions in Nature and Society* (Princeton, NJ: Princeton University Press, 2009), 169-70.

Image Credits

Figure 1. Paul K. Chien in front of Maotian Shan. Photograph by Illustra Media. Used with permission.

Figures 2–3. Stellostomites and Hyoliths. Photographs by Paul K. Chien.

Figure 4. Maotianshania cylindrica. Photograph by Illustra Media. Used with permission.

Figures 5–6. Leanchoilia and trilobite. Photograph by Illustra Media. Used with permission.

Figure 7. Haikouella. Photograph by Paul K. Chien.

Figure 8. Phyla graphic. Recreation by Eric H. Anderson, based on information provided by D. G. Shu to the author.

Figure 9. Graphical representation of the "Hard Facts Wall" and actual data. Image by Access Research Network. Used with permission.

Figure 10. California Academy of Sciences museum display of the "Timeline of Life on Earth." Photograph by Paul K. Chien.

Figure 11. Panel from exhibit at the Beijing National Museum of Natural History. Photograph by Paul K. Chien.

Figure 12. Small round fossils. Photograph by Paul K. Chien.

Figure 13. Sponge egg images. Photographs by Paul K. Chien.

WHAT IS THE DISCOVERY SOCIETY?

The Discovery Society is a group of individuals who come together to support the work—and disseminate the message—of Discovery Institute's Center for Science and Culture. New members receive materials that help educate themselves and spread the word about our work to those in their circle of influence. Depending upon their giving level, members receive one to three Discovery Institute Press newly released books per year, along with invitations to regional donor events and discounted rates on our annual Insiders Briefing events.

If you appreciate this booklet and aren't already a member, we hope you will consider joining our network of supporters today!

Your donation to Discovery Institute's Center for Science and Culture will allow us to expand our cutting-edge scientific research and scholarship; train young people through our education and outreach; and reach the masses through media and communications.

discovery.org/id/donate

MORE INFORMATION ON THE DISCOVERY SOCIETY CAN BE FOUND AT
discovery.org/id/donate/#member-levels.

EVOLUTION AND INTELLIGENT DESIGN IN A NUTSHELL

Are life and the universe a mindless accident—the blind outworking of laws governing cosmic, chemical, and biological evolution? That's the official story many of us were taught somewhere along the way. But what does the science actually say? Drawing on recent discoveries in astronomy, cosmology, chemistry, biology, and paleontology, *Evolution and Intelligent Design in a Nutshell* shows how the latest scientific evidence suggests a very different story.

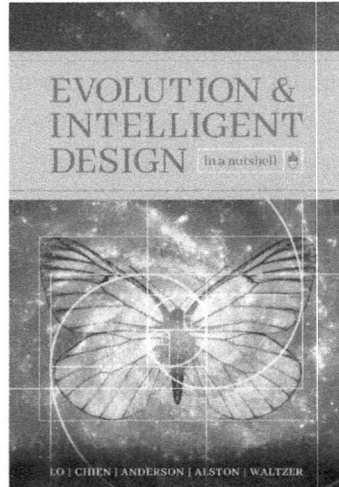

EVOLUTION & INTELLIGENT DESIGN in a nutshell

LO | CHEN | ANDERSON | ALSTON | WALTZER

"accessible, informative… powerful … an excellent resource."

J. Warner Wallace

PURCHASE THE FULL BOOK HERE:

DiscoveryInstitutePress.com/EvolutionandID

More in This Series:

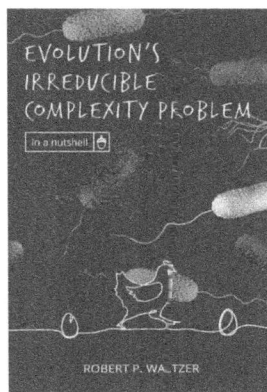

THE BIG BANG &
THE FINE-TUNED
UNIVERSE
in a nutshell
ROBERT A. ALSTON

THE ORIGIN OF LIFE
& THE INFORMATION
PROBLEM
in a nutshell
ERIC H. ANDERSON

FACTORIES THAT
BUILD FACTORIES
in a nutshell
ERIC H. ANDERSON

EVOLUTION'S
IRREDUCIBLE
COMPLEXITY PROBLEM
in a nutshell
ROBERT P. WALTZER

This series of booklets was created to help Discovery Society members educate themselves about the basic arguments for intelligent design and the critiques of Darwinian evolution. Each booklet presents the content of one chapter of *Evolution and Intelligent Design in a Nutshell*. To help you delve deeper into each subject, we have included a list of recommended resources from our vast library of videos, podcasts, articles, and websites. Members of the Discovery Society can download digital versions of these books through the Discovery Society Community on the DiscoveryU.org platform or purchase physical copies at a discounted rate through Amazon.com.

www.ingramcontent.com/pod-product-compliance
Lightning Source LLC
Chambersburg PA
CBHW022100210326
41520CB00046B/795